On Sum of Divisors and Robin's Theorem

Valentin B. Bura

1. Introduction

The focus of this paper is the inequality between the sum of divisors function $\sigma(n)$ and the expression $e^\gamma n \log \log n$.

The function $\sigma(n)$ gives the sum of all divisors for n. A function related to $\sigma(n)$ is the divisor function $d(n)$ which provides the number of divisors for n.

The Riemann Hypothesis was proposed in 1859 in [2]. The connection between the divisor function σ and the Riemann Hypothesis was studied by Robin in an article written in 1984 [1].

Erdős and Alaoglu studied in 1944 [3] the relationship between this function and highly composite numbers, a notion defined by Srinivasa Ramanujan in 1915 [6].

The functions $\sigma(n)$ and $d(n)$ were recurring themes for Paul Erdős. He was trying to identify upper and lower bounds on these functions by considering special values for n [12].

Ramanujan was the first to prove also in [6] the Riemann Hypothesis implies for large n the inequality $\sigma(n) < e^\gamma n \log \log n$.

In 1984, Robin proved that this inequality is true for all $n > 5040$ if and only if the Riemann Hypothesis holds [1]. This result is referred to as Robin's Theorem.

In this paper we are concerned with this condition that falsifies the Riemann Hypothesis, derived from Robin's Theorem.

The condition that falsifies the Riemann Hypothesis is the fluctuation in values taken by the sum of divisors function above the given threshold $e^\gamma n \log \log n$ for infinitely many n.

We are able to show this condition does not hold for infinitely many numbers. We achieve this by making use of two exact expressions for sum of divisors in terms of factorizations into primes.

These expressions being exact it is then somewhat straightforward to show by induction that $\sigma(n)$ is less than $e^\gamma n \log \log n$ for values of n sufficiently large.

Given a positive integer m, the type of induction we perform is on the length k of the prime factorization $m = b_1 \ldots b_k$.

We begin as usual with $k = 1$, in which case m itself is an arbitrary prime, and for an arbitrary value $m = t$, we consider the value $t \times b$ for any prime b.

The unique factorization theorem ensures that any m can be represented as the product of prime numbers and this representation is unique up to order of factors.

Hence there exists a bijection between positive integers greater than 1 and such factorizations into primes. This bijection ensures that our induction method, apart from being sound, is also complete.

Analyses of the values of $d(n)$ and $\sigma(n)$ have focused on the usual order on the positive integers induced by the successor function $n \mapsto n + 1$.

An analysis involving $\sigma(n)$ is found in [7], where an expression is derived for a Dirichlet series involving $\sigma(n)$.

An early analysis of $d(n)$ is the one in [8] which considers the question posed by Erdős and Mirsky in [4] whether there exist infinitely many integers n for which $d(n) = d(n + 1)$.

More recent analyses are as follows. In [9] the authors study the symmetry of the divisor function over some specific intervals. In [10] the authors derive lower and upper bounds for the average value of the ratio of consecutive values of the function. In [11] a correlation is presented between n and $d(n)$.

We study here the values of $\sigma(n)$, and Robin's inequality $\sigma(n) < e^\gamma n \log \log n$.

Our analysis assumes the partial order on the integers induced by each integer's prime factorization. The partial order is the following.

$$R(\emptyset, b) \text{ for prime } b$$

$$R(b_1, b_2) \text{ for primes } b_1, b_2$$

$$R(n, m) \wedge R(q, b) \to R(nmq, b) \text{ for positive integers } n, m, q \text{ and prime } b$$

Since any positive integer m greater than 1 admits such a prime factorization, we have for each such m with factorization $m = b_1 b_2 \ldots b_k b_{k+1}$ the following corresponding sequence.

$$R(\emptyset, b_1), R(b_1, b_2), \ldots, R(b_1 b_2 \ldots b_k, b_{k+1})$$

This is the gist of our method and induction is performed on this sequence.

We make use of this method to show that for any positive integer n greater than a certain threshold, if $n = m \times q$ for q a prime or a power of a prime, we have $\sigma(n) < e^\gamma n \log \log n$.

We show this by arguing that $n = m \times q$ where q is either a prime or a power of a prime greater than 11, and m is a positive integer.

The condition on q is that it is either a power of two greater than four, or a power of three greater than two, or a power of five greater than two, or a power of seven greater than one, or a prime or a power of a prime greater than ten.

In particular, we show that if $n \geq 2^5 \times 3^3 \times 5^3 \times 7^2$ then $\sigma(n) < e^\gamma n \log \log n$.

2. Preliminaries

In this work we only refer to non-negative elements of $\mathbb{Z}, \mathbb{Q}, \mathbb{R}$. Let $\mathbb{N}^* = \mathbb{N} \setminus \{0\}$. We refer to $\mathbb{N}^*$ as the set of positive integers. If a real x evaluates to a positive integer we say x has integral value.

We use i, j, k, s for indexing. We use m, n, q to denote positive integers. We use $a, a_1, a_2 \ldots a_s$ to denote integral exponents of powers. We use $p_1, p_2 \ldots p_s$ and $b_1, b_2 \ldots b_k$ to denote prime numbers. We use f, g to denote functions.

For positive integers m and n we say n is a divisor of m and write $n \mid m$ if there exists positive integer c such that $m = n \times c$. If $c \neq 1$ and $c \neq m$ we say n is a proper divisor of m.

For positive integers m and n denote explicit fractions or division as m/n.

A function f is decreasing if $m < n$ implies $f(n) < f(m)$.

A function f is increasing if $m < n$ implies $f(m) < f(n)$.

For a finite series of the form $1 + r + r^2 + \cdots + r^n$ the summation is the following.

$$\sum_{i=0}^{n} r^i = \frac{r^{n+1} - 1}{r - 1}$$

The function $f(n) = q - 1/n$ for constant q converges asymptotically to q since we have the following.

$$\lim_{n \to \infty} f(n) = \lim_{n \to \infty} (q - 1/n) = q - \lim_{n \to \infty} 1/n = q$$

The function $g(n) = \log \log n$ is increasing since we have the following.

$$g(n) < g(m) \text{ for } n < m, \text{ that is } \log \log n < \log \log m$$

Remark 1 (Unique Factorization Theorem). *Every integer greater than 1 is either a prime number itself, or can be represented as the product of prime numbers. This representation is unique up to the order of factors.*

We let p_n be the nth prime. For any natural number m commence with its compact factorization into primes.

$$m = p_1^{a_1} \ldots p_s^{a_s}$$

This factorization into primes has the extended form $m = b_1 \ldots b_k$. Each b_i is a (not necessarily distinct) prime and the following holds.

$$k = \sum_{i=1}^{s} a_i$$

Definition 2. *Define $\sigma(m) = \sum_{p \mid m} p$.*

Definition 3. *Define $\Omega = 2^5 \times 3^3 \times 5^3 \times 7^2$.*

We define the constant δ. using the values provided in [1] for γ and e^γ.

$$\gamma = 0.577215\ldots \text{ and } e^\gamma = 1.78107\ldots$$

Definition 4. *Let $\delta = 1781 \times 10^{-3}$.*

Remark 5. *Note that $\delta x < e^\gamma x$ for any $x \in \mathbb{R}$ with $x > 0$.*

We define symbolic notations ϕ and φ.

Definition 6. *Define $\phi(m) = \delta \times m \times \log\log m$.*

Definition 7. *Define $\varphi(m) = \delta \times \log\log m$.*

Note $\varphi(m) = \phi(m)/m$.

We define maps τ and ξ.

Definition 8. *Let p be a prime. Define $\tau(p, a) = \sum_{i=0}^{a} p^i$.*

Definition 9. *Suppose m is written as $m = p_1^{a_1} \ldots p_s^{a_s}$. Define $\xi(p, a) = \sum_{i=0}^{a} 1/p^i$. Define $\xi(m) = \prod_{i=1}^{k} \xi(p_i, a_i)$.*

We define maps α, β.

Definition 10. *Let $m = b^a$ for a prime b. Define $\alpha(m)$ as follows.*

$\alpha(m) = b^5$ *for $b = 2$.*

$\alpha(m) = b^3$ *for $b = 3$.*

$\alpha(m) = b^3$ *for $b = 5$.*

$\alpha(m) = b^2$ *for $b = 7$.*

$\alpha(m) = b$ *for $b \geq 11$.*

Definition 11. *Let m be a power of a prime. Let $\beta(m) = m/\alpha(m)$.*

We give two remarks on the connection between R.H. and the divisor function.

Remark 12. *Robin in [1] shows the R.H. implies the following for $n \geq 5041$.*

$$\sigma(m) < e^\gamma m \log\log(m)$$

Remark 13. *Robin in [1] shows that if the R.H. is false, the following holds for infinitely many n.*

$$\sigma(m) > e^\gamma m \log\log(m)$$

In the light of the two Remarks above, and considering Remark 13 in particular, it is sufficient to show that $\sigma(m) < e^\gamma m \log\log(m)$ for only finitely many m in order to settle the Riemann Hypothesis.

This key observation outlines the theoretical foundation for our investigation.

For $n < 5041$ the positive integers for which $\sigma(m) > e^\gamma m \log\log(m)$ are 2, 3, 4, 5, 6, 8, 9, 10, 12, 16, 18, 20, 24, 30, 36, 48, 60, 72, 84, 120, 180, 240, 360, 720, 840, 2520, 5040.

For certain combinations of the following exponents,

$$a_1 = 0, 1, 2, 3, 4, a_2 = 0, 1, 2, a_3 = 0, 1, 2, a_4 = 0, 1,$$

the prime factorizations for these positive integers are of the following form.

$$2^{a_1} \times 3^{a_2} \times 5^{a_3} \times 7^{a_4}$$

3. Initial considerations

The following expression for σ is given by Hardy and Wright in [5]. For brevity of exposition we assume this expression without proof.

Remark 14. *For $m = p_1^{a_1} \ldots p_k^{a_k}$ the values of $\sigma(m)$ are as follows.*

$$\sigma(m) = \prod_{i=1}^{k} \sum_{j=0}^{a_i} p_i^j$$

$$\therefore \sigma(m) = \prod_{i=1}^{k} \tau(p_i, a_i)$$

We define the map Λ. In the next section we show that Λ and σ are the same.

Definition 15. *For $m = p_1^{a_1} \ldots p_k^{a_k}$ define the map $\Lambda(m)$ as follows.*

Let $\lambda(m) = m$. Let $\lambda(m \setminus p_i) = m/p_i^{a_i}$. Note in the case $p_i^{a_i} \mid m$, $\lambda(m \setminus p_i)$ is of integral value.

For a prime b and any composite number m define Λ.

$$\Lambda(1) = 1$$

$$\Lambda(mb) = \begin{cases} b \times \Lambda(m) + \Lambda(\lambda(m \setminus b)) & b \mid m \\ b \times \Lambda(m) + \Lambda(m) & b \nmid m. \end{cases}$$

We provide this piecewise definition to give transparency to our proofs.

Note that a more concise definition for Λ in the inductive step is the following.

$$\Lambda(mb) = b \times \Lambda(m) + \Lambda(\lambda(m \setminus b))$$

This is due to $\lambda(m \setminus b) = m$ in the case $b \nmid m$.

Lemma 16. and Lemma 17. are concerned with divisibility by a prime. These Lemmas are used further in the proofs of Lemma 18. and Lemma 30.

Lemma 16. *Let $b, b_1, \ldots b_k$ be prime numbers. Let $b \mid (b_1 \ldots b_k)$.*

Then $b = b_i$ for some i with $1 \leq i \leq k$.

Proof

Proceed by induction on k.

If $k = 1$ we have $b \mid b_1$. Since b_1 is prime, it has no factors other 1 and b_1. So b is either 1 or b_1. However b is a prime so it cannot be 1. Therefore $b = b_1$. This concludes the base case.

Take some arbitrary k and suppose the statement holds for all values up to and including k. We have the following.

$$b \mid (b_1 \ldots b_k) \text{ implies } m = b_i \text{ for some } i \text{ with } 1 \leq i \leq k$$

Consider $k + 1$. We have $m \mid b_1 \ldots b_k b_{k+1}$. We rearrange this expression.

$$b \mid ((b_1 \ldots b_k) b_{k+1})$$

Write $n = b_1 \ldots b_k$. Therefore $b \mid n \times b_{k+1}$. If $b \mid n$ by inductive hypothesis we have $b = b_i$ for some i with $1 \leq i \leq k$. If $b \mid b_{k+1}$ then by an argument similar to the argument in the base step we have $b = b_k$.

If $b \mid n \times p_1 \times \cdots \times p_s$ for $\{p_1, \ldots, p_s\} \subseteq \{b_1, \ldots, b_k\}$ then the implication reduces to a previous value of k.

This concludes the proof of Lemma 16.

□

Lemma 17. *Let $b, b_1, \ldots b_k$ be prime numbers. Let $b \nmid (b_1 \ldots b_k)$.*

Then $b \neq b_i$ for any i with $1 \leq i \leq k$.

Proof

Suppose that $b = b_i$ for some i with $1 \leq i \leq k$.

Then $b \mid (b_1 \ldots b_k)$, a contradiction.

$\square$

Lemma 18. *Let $m \geq \Omega$. Then at least one of the following holds.*

1. *$2^a \mid m$ for $a \geq 5$,*

2. *$3^a \mid m$ for $a \geq 3$,*

3. *$5^a \mid m$ for $a \geq 3$,*

4. *$7^a \mid m$ for $a \geq 2$.*

5. *$b^a \mid m$ for prime $b \geq 11$ and $a \geq 1$,*

Proof

Suppose for a contradiction none of the conditions hold.

Write $m = b_1 \ldots b_k$.

Since condition 5. does not hold we have from Lemma 16. that all b_i in the prime factorization of m are one of the values $2, 3, 5, 7$.

Since conditions 1. 2. 3. 4. do not hold, we have that $m \leq 2^4 \times 3^2 \times 5^2 \times 7^2$.

Therefore $m < \Omega$. A contradiction.

This concludes the proof of Lemma 18.

$\square$

Lemma 19. states that the function φ is increasing. Lemma 20. states that the function $1 + 1/n$ is decreasing.

Lemma 19. *Let $m < n$. Then $\varphi(m) < \varphi(n)$.*

Proof

Suppose $m < n$.

The function $f(n) = \log \log n$ is increasing.

The following are equivalent.

$$m < n$$

$$\log \log m < \log \log n$$

$$\delta \log \log m < \delta \log \log n$$

$$\varphi(m) < \varphi(n)$$

This concludes the proof of Lemma 19.

□

Lemma 20. *Let $m < n$. Then $1 + 1/n < 1 + 1/m$.*

Proof

If $m < n$ then $1/n < 1/m$ then $1 + 1/n < 1 + 1/m$

□

Lemma 21. is concerned with Robin's inequality for primes greater than 11.

Lemma 21. *Let p be a prime with $p \geq 11$. Then $\sigma(p) < \phi(p)$.*

Proof

We need to show $\sigma(p) < \phi(p)$. If p is a prime then $\sigma(p) = p+1$. The following are therefore equivalent.

$$\sigma(p) < \phi(p)$$

$$p + 1 < \phi(p)$$

$$p + 1 < \delta p \log \log p$$

$$1 + 1/p < \delta \log \log p$$

We observe the following.

$$1 + 1/11 < 13/10 < 15/10 < \delta \log \log 11$$

$$\therefore 1 + 1/11 < \delta \log \log 11$$

Using Lemma 20, the function $f(n) = 1 + 1/n$ is decreasing, hence for $n \geq 11$ we have $f(11) > f(n)$. Using Lemma 19, the function $\varphi(n)$ is strictly increasing. Therefore for $p \geq 11$ we observe the following.

$$(1 + 1/p) < (1 + 1/11) < \delta \log \log 11 < \delta \log \log m$$

$$\therefore (1 + 1/p) < \varphi(m)$$

$$\therefore \sigma(p) < \phi(p)$$

This concludes the proof of Lemma 21.

□

Lemma 22. is concerned with a lower bound for $\delta \log \log 2^a$.

Lemma 22. *Let $q = 2^a$ for $a \geq 5$. We have $2 < \varphi(q)$.*

Proof

We need to show $2 < \varphi(q)$.

So we need to show $2 < \varphi(2^a)$ for $a \geq 5$.

If $a = 5$ we have the following.

$$\varphi(2^a) = \varphi(2^5) = \delta \log \log 2^5$$

And we have the following.

$$2 < 22/10 < \delta \log \log 2^5$$

$$\therefore 2 < 22/10 < \varphi(2^5)$$

$$\therefore 2 < \varphi(2^5)$$

Using Lemma 19, the function $\varphi(n)$ is increasing.

Therefore we have the following for $a \geq 5$.

$$2 < \varphi(2^5) \leq \varphi(2^a)$$

$$\therefore 2 < \varphi(q)$$

This concludes the proof of Lemma 22.

□

Lemma 23, is concerned with a lower bound for $\delta \log \log 3^a$.

Lemma 23. *Let $q = 3^a$ for $a \geq 3$. We have $3/2 < \varphi(q)$.*

Proof

We need to show $3/2 < \varphi(q)$.

So we need to show $3/2 < \varphi(3^a)$ for $a \geq 3$.

If $a = 3$ we have the following.

$$\varphi(3^a) = \varphi(3^3) = \delta \log \log 3^3$$

And we have the following.

$$3/2 < 21/10 < \delta \log \log 3^3$$

$$\therefore 3/2 < 21/10 < \varphi(3^3)$$

$$\therefore 3/2 < \varphi(3^3)$$

Using Lemma 19, the function $\varphi(n)$ is increasing.

Therefore we have the following for $a \geq 3$.

$$3/2 < \varphi(3^3) \leq \varphi(3^a)$$

$$\therefore 3/2 < \varphi(q)$$

This concludes the proof of Lemma 23.

Lemma 24. is concerned with a lower bound for $\delta \log \log 5^a$.

Lemma 24. *Let $q = 5^a$ for $a \geq 3$. We have $5/4 < \varphi(q)$.*

Proof

We need to show $5/4 < \varphi(q)$.

So we need to show $5/4 < \varphi(5^a)$ for $a \geq 3$.

If $a = 3$ we have the following.

$$\varphi(5^a) = \varphi(5^3) = \delta \log \log 5^3$$

And we have the following.

$$5/4 < 28/10 < \delta \log \log 5^3$$

$$\therefore 5/4 < 28/10 < \varphi(5^3)$$

$$\therefore 5/4 < \varphi(5^3)$$

Using Lemma 19. the function $\varphi(n)$ is increasing.

Therefore we have the following for $a \geq 3$.

$$5/4 < \varphi(5^3) \leq \varphi(5^a)$$

$$\therefore 5/4 < \varphi(q)$$

This concludes the proof of Lemma 24.

Lemma 25. is concerned with a lower bound for $\delta \log \log 7^a$.

Lemma 25. *Let $q = 7^a$ for $a \geq 2$. We have $7/6 < \varphi(q)$.*

Proof

We need to show $7/6 < \varphi(q)$.

So we need to show $7/6 < \varphi(7^a)$ for $a \geq 2$.

If $a = 2$ we have the following.

$$\varphi(7^a) = \varphi(7^2) = \delta \log \log 7^2$$

And we have the following.

$$7/6 < 24/10 < \delta \log \log 7^2$$

$$\therefore 7/6 < 24/10 < \varphi(7^2)$$

$$\therefore 7/6 < \varphi(7^2)$$

Using Lemma 19. the function $\varphi(n)$ is increasing.

Therefore we have the following for $a \geq 3$.

$$7/6 < \varphi(7^2) \leq \varphi(7^a)$$

$$\therefore 7/6 < \varphi(q)$$

This concludes the proof of Lemma 25.

□

Lemma 26. *Let $q = b^a$ for prime $b \geq 11$ and $a \geq 1$. We have $11/10 < \varphi(q)$.*

Proof

We need to show $11/10 < \varphi(q)$.

So we need to show $11/10 < \varphi(11^a)$ for $a \geq 5$.

If $a = 1$ we have the following.

$$\varphi(11^a) = \varphi(11) = \delta \log \log 11$$

And we have the following.

$$11/10 < 16/10 < \delta \log \log 11$$

$$\therefore 11/10 < 16/10 < \varphi(11)$$

$$\therefore 11/10 < \varphi(11)$$

Using Lemma 19, the function $\varphi(n)$ is strictly increasing with n.

Therefore we have the following for $a \geq 1$.

$$11/10 < \varphi(11) \leq \varphi(11^a)$$

Therefore we have the following for $b \geq 11$ and $a \geq 1$.

$$11/10 < \varphi(b^a)$$

$$\therefore 11/10 < \varphi(q)$$

This concludes the proof of Lemma 26.

□

4. Inequalities and equivalences

Results in this section are Lemmas 27, 28, 29, 30, 31, 32, 33, 34, 35, 36, 37.

Lemma 27 gives an equivalence involving sum of divisors for powers of primes.

Lemma 28 gives a relation between the maps τ and ξ.

Lemma 29 provides the inequality $(q+1)\phi(m) < \phi(mq)$ for $m, q \geq 11$.

The main result of this section is Lemma 30. This proves the identification between the functions σ and Λ.

Hence we show for a prime b that $\sigma(mb) = (b+1)\sigma(m)$ in the case $b \nmid m$, and $\sigma(mb) = b\sigma(m) + \sigma(\lambda(m \setminus b))$ in the case $b \mid m$.

Lemma 31. states that $(b+1)\sigma(m) < \phi(mb)$ for an odd prime b and $m \geq 11$.

Lemma 32. and Lemma 33. give an expression and an inequality involving the function Λ.

Lemma 34. gives the inequality $\sigma(mb) < (b+1)\sigma(m)$.

Lemma 35. gives a simple expression involving $\sigma(m)$ and $\xi(m)$.

Lemma 36. gives equivalences involving $\sigma(m)$, $\phi(m)$, $\varphi(m)$ and $\xi(m)$.

Lemma 37. gives an inequality involving the map ξ.

We prove the equivalence between the sum of divisors for powers of primes and the function τ.

Lemma 27. *Let b be a prime. Let $q = b^a$ for $a \geq 1$.*

Then we observe the following.

$$\sigma(q) = \tau(b, a)$$

Proof

We need to show $\sigma(q) = \tau(b, a)$.

So we need to show $\sigma(b^a) = \tau(b, a)$ for $a \geq 1$. Proceed by induction on a.

If $a = 1$ since b is a prime we have the following.

$$\sigma(b) = b + 1 = \tau(b, 1)$$

$$\therefore \sigma(b^a) = \tau(b, a)$$

Suppose for some arbitrary $a \geq 1$ the following statement holds.

$$\sigma(b^a) = \tau(b, a)$$

Consider the case of $a + 1$. The following are equivalent.

$$\sigma(b^{a+1}) = \tau(b, a+1)$$

$$\Lambda(b^{a+1}) = \tau(b, a+1)$$

$$b\Lambda(b^a) + \Lambda(\lambda(b^a \setminus b)) = b\tau(b, a) + 1$$

$$b\Lambda(b^a) + \Lambda(1) = b\tau(b, a) + 1$$

$$b\Lambda(b^a) + 1 = b(b^a + b^{a-1} + \cdots + b + 1) + 1$$

$$b\Lambda(b^a) + 1 = b^{a+1} + b^a + \cdots + b^2 + b + 1$$

$$b\Lambda(b^a) + 1 = \tau(b^{a+1})$$

The following are equivalent.

$$\Lambda(b^a b) = \sigma(b^a b)$$

$$\Lambda(b^a b) = \sigma(b^{a+1})$$

$$b\Lambda(b^a) + \Lambda(\lambda(b^a \setminus b)) = \sigma(b^{a+1})$$

$$b\Lambda(b^a) + \Lambda(1) = \sigma(b^{a+1})$$

$$b\Lambda(b^a) + 1 = \sigma(b^{a+1})$$

$$\tau(b, a+1) = \sigma(b^{a+1})$$

Therefore $\sigma(b, a+1) = \tau(b, a+1)$.

By induction we have $\sigma(b^a) = \tau(b, a)$. Therefore $\sigma(q) = \tau(b, a)$.

This concludes the proof of Lemma 27.

□

We give an expression that relates the functions τ and ξ.

Lemma 28. *For $n \geq 1$ and $a \geq 1$ we observe the following.*

$$\frac{\tau(n,a)}{n^a} = \xi(n,a)$$

Proof

We need to show the following.

$$\frac{\tau(n,a)}{n^a} = \xi(n,a)$$

We have the following.

$$\xi(n,a) = 1 + 1/n + 1/n^2 + \cdots + 1/n^a$$

$$\tau(n,a) = n^a + n^{a-1} + \cdots + n + 1$$

$$\therefore \frac{\tau(n,a)}{n^a} = \frac{n^a + n^{a-1} + \cdots + n + 1}{n^a}$$

$$\therefore \frac{\tau(n,a)}{n^a} = 1 + 1/n + 1/n^2 + \cdots + 1/n^a$$

$$\therefore \frac{\tau(n,a)}{n^a} = \xi(n,a)$$

This concludes the proof of Lemma 28.

□

Lemma 29. is a technical requirement. This Lemma is further used for proving the results of Lemmas 48, 49, 50, 51, 52, 53.

Lemma 29. *Let $m \geq 11$ and let $q \geq 11$. we have $(q+1)\phi(m) < \phi(mq)$.*

Proof

We need to show $(q+1)\phi(m) < \phi(mq)$.

The following are equivalent.

$$(q+1)\phi(m) < \phi(mq)$$

$$(q+1)\delta m \log\log m < \delta mb \log\log mq$$

$$(q+1)\log\log m < q\log\log mq$$

$$(1+1/q)\log\log m < \log\log mq$$

We observe the following numerical inequality.

$$(1+1/11)\log\log 11 < 15/10 < \log\log 11 \times 11$$

$$\therefore (1+1/11)\log\log 11 < \log\log 11 \times 11$$

Using Lemma 20, the function $f(n) = 1 + 1/n$ is decreasing.

We can see the following function is increasing for constant m and $q \geq 11$.

$$g(m,q) = \frac{\log\log mq}{\log\log m}$$

$$\therefore f(q) < g(m,q)$$

Therefore we have the following for $m \geq 11$ and $q \geq 11$.

$$f(q) < g(m,q)$$

$$\therefore 1 + 1/q < \frac{\log\log mq}{\log\log m}$$

$$\therefore (1 + 1/q)\log\log m < \log\log mq$$

$$\therefore (q+1)\phi(m) < \phi(mq)$$

This concludes the proof of Lemma 29.

Lemma 30. *For all $m \geq 2$ we observe the following.*

$$\sigma(m) = \Lambda(m)$$

Proof

Write $m = b_1 \ldots b_k$ and compactly $m = p_1^{a_1} \ldots p_s^{a_s}$. From Remark 14, we have the following expression for $\sigma(m)$.

$$\sigma(m) = \prod_{i=1}^{s} \tau(p_i, a_i)$$

We show the following.

$$\Lambda(m) = \prod_{i=1}^{k} \tau(p_i, a_i)$$

Proceed by induction on k.

If $k = 1$ let $m = b$. We have the following.

$$\Lambda(m) = b \times \Lambda(1) + \Lambda(\lambda(m \setminus b))$$

$$\Lambda(m) - b + \Lambda(1) - b + 1$$

$$\therefore \Lambda(b) = \sigma(b)$$

Suppose for some k we have the following.

$$\Lambda(b_1 \ldots b_k) = \sigma(b_1 \ldots b_k)$$

Consider the case of $k + 1$.

Suppose $b_{k+1} \nmid (b_1 \ldots b_k)$. We have the following.

$$\Lambda(b_1 \ldots b_k b_{k+1}) = (b_{k+1} + 1)\Lambda(b_1 \ldots b_k) = (b_{k+1} + 1)\sigma(b_1 \ldots b_k)$$

And we observe the following.

$$(b_{k+1} + 1)\sigma(b_1 \ldots b_k) = (b_{k+1} + 1)\prod_{i=1}^{s} \tau(p_i, a_i)$$

We have from Lemma 17 that $b_{k+1} \neq p_i$ for $1 \leq i \leq s$. Therefore for $p_{s+1} = b_{k+1}$ and $a_{s+1} = 1$ we have the following.

$$(b_{k+1} + 1)\sigma(b_1 \ldots b_k) = \prod_{i=1}^{s+1} \tau(p_i, a_i)$$

$$\therefore \Lambda(b_1 \ldots b_k b_{k+1}) = \sigma(b_1 \ldots b_k b_{k+1})$$

Suppose $b_{k+1} \mid (b_1 \ldots b_k)$. We have the following.

$$\Lambda(b_1 \ldots b_k b_{k+1}) = b_{k+1}\Lambda(b_1 \ldots b_k) + \Lambda(\lambda(b_1 \ldots b_k \setminus b_{k+1}))$$

And we observe the following.

$$(b_{k+1} + 1)\sigma(b_1 \ldots b_k) = (b_{k+1} + 1)\prod_{i=1}^{s} \tau(p_i, a_i)$$

Therefore for $b_{k+1} \in \{p_1, p_2 \ldots p_s\}$ and $c_i = a_i$ for $i \neq k$ while $c_{k+1} = a_i + 1$ we have the following.

$$(b_{k+1} + 1)\sigma(b_1 \ldots b_k) = \prod_{i=1}^{s} \tau(p_i, c_i)$$

We expand Λ and observe the following.

$$\Lambda(b_1 \ldots b_k b_{k+1}) = b_{k+1}\Lambda(b_1 \ldots b_k) + \Lambda(\lambda(b_1 \ldots b_k \setminus b_{k+1}))$$

$$\therefore \Lambda(b_1 \ldots b_k b_{k+1}) = b_{k+1}\sigma(b_1 \ldots b_k) + \Lambda(\lambda(b_1 \ldots b_k \setminus b_{k+1}))$$

Suppose we have the following.

$$\Lambda(\lambda(b_1 \ldots b_k \setminus b_{k+1})) = \prod_{i=1}^{s} \tau(p_i, a_i)$$

Then the following holds.

$$b_{k+1}\Lambda(b_1 \ldots b_k) = \prod_{i=1}^{s} \tau(p_i, a_i) + \tau(b_{k+1}, a_{k+1})$$

$$\therefore \Lambda(b_1 \ldots b_k b_{k+1}) = \prod_{i=1}^{s+1} \tau(p_i, a_i)$$

$$\therefore \Lambda(b_1 \ldots b_k b_{k+1}) = \sigma(b_1 \ldots b_k b_{k+1})$$

By induction we have $\sigma(m) = \Lambda(m)$.

This concludes the proof of Lemma 30.

Lemma 31. *Let m be a positive integer with $m \geq 11$. Suppose $\sigma(m) < \phi(m)$. Let b be an odd prime. Then $(b+1)\sigma(m) < \phi(mb)$.*

Proof

The following are equivalent.

$$\sigma(m) < \phi(m)$$

$$(b+1)\sigma(m) < (b+1)\phi(m)$$

$$(b+1)\sigma(m) < (b+1)\delta m \log \log m$$

The following are equivalent.

$$(b+1)\delta m \log \log m < \delta mb \log \log mb$$

$$(b+1)\log \log m < b \log \log mb$$

$$(1 + 1/b)\log \log m < \log \log mb$$

We have the following numerical inequalities.

$$4/3 \log \log 11 < \log \log 3 \times 11$$

$$\therefore (1 + 1/3)\log \log 11 < \log \log 3 \times 11$$

Using Lemma 20, the function $f(n) = 1 + 1/n$ is decreasing, hence for $n \geq 3$ we have $f(3) > f(n)$.

Therefore we observe the following.

$$\therefore (1 + 1/b) \log \log m \leq (1 + 1/3) \log \log m$$

$$\therefore (1 + 1/b) \log \log m < \log \log mb$$

$$\therefore (b+1)\delta m \log \log m < \delta mb \log \log mb$$

$$\therefore (b+1)\sigma(m) < \phi(mb)$$

This concludes the proof of Lemma 31.

□

From Lemma 30., Remark 14, and Definition 15, we obtain Lemma 32.

Lemma 32. *Let $m = p_1^{a_1} \ldots p_s^{a_s}$. Then we have the following.*

$$\Lambda(\lambda(m \setminus p_j)) = \prod_{i=1}^{j-1} \tau(p_i, a_i) + \prod_{i=j+1}^{s} \tau(p_i, a_i)$$

Using Lemma 32, we obtain Lemma 33.

Lemma 33. *For all $m \geq 2$ and prime b we observe $\Lambda(\lambda(m \setminus b)) \leq \Lambda(m)$.*

Lemma 34. *Let m be a positive integer, and let b be an odd prime. Suppose $\sigma(m) < \phi(m)$. Then $\sigma(mb) \leq (b+1)\sigma(m)$.*

Proof

We have $\sigma(m) = \Lambda(m)$.

If $b \nmid m$ we have the following.

$$\Lambda(mb) = (b+1)\Lambda(m)$$

$$\therefore \sigma(mb) = (b+1)\sigma(m)$$

If $b \mid m$ we have the following.

$$\Lambda(mb) = b\Lambda(m) + \Lambda(\lambda(m \setminus b))$$

Using Lemma 33, we have

$$\Lambda(\lambda(m \setminus b)) \leq \Lambda(m)$$

$$\therefore \Lambda(mb) \leq b\Lambda(m) + \Lambda(m)$$

$$\therefore \Lambda(mb) \leq (b+1)\Lambda(m)$$

$$\therefore \sigma(mb) \leq (b+1)\sigma(m)$$

This concludes the proof of Lemma 34.

□

Lemma 35. *Let $m \geq 2$ and let $m = p_1^{a_1} \ldots p_k^{a_k}$. We observe the following.*

$$\frac{\sigma(m)}{m} = \xi(m)$$

Proof

We have the following.

$$\frac{\sigma(m)}{m} = \frac{\prod_{i=1}^{k} \sum_{j=0}^{a_i} p_i^j}{\prod_{i=1}^{k} p_i^{a_i}}$$

$$\therefore \frac{\sigma(m)}{m} = \prod_{i=1}^{k} \frac{\sum_{j=0}^{a_i} p_i^j}{p_i^{a_i}} = \prod_{i=1}^{k} \frac{1 + p_i + \cdots + p_i^{a_i}}{p_i^{a_i}}$$

$$\therefore \frac{\sigma(m)}{m} = \prod_{i=1}^{k} (\frac{1}{p_i^{a_i}} + \frac{1}{p_i^{a_i-1}} + \cdots + 1)$$

$$\therefore \frac{\sigma(m)}{m} = \prod_{i=1}^{k} \xi(p_i, a_i) = \xi(m)$$

□

Lemma 36. *The following are equivalent.*

$$\sigma(m) < \phi(m)$$
$$\xi(m) < \varphi(m)$$

Proof

Using Lemma 35, the following are equivalent.

$$\sigma(m) < \phi(m)$$

$$\sigma(m) < \delta m \log \log m$$

$$\frac{\sigma(m)}{m} < \delta \log \log m$$

$$\xi(m) < \varphi(m)$$

□

Lemma 37. *For any finite a and $p > 1$ we observe the following.*

$$\xi(p, a) < \frac{p}{p-1}$$

Proof

We have the following.

$$\xi(p, a) = \sum_{i=0}^{a} 1/p^i$$

$$\therefore \xi(p, a) = 1 + 1/p + 1/p^2 + \cdots + 1/p^a$$

A finite series of the form $1 + r + r^2 + \cdots + r^a$ with $r = 1/p$.

$$\therefore \xi(p, a) = \frac{1/p^{a+1} - 1}{1/p - 1}$$

$$\therefore \xi(p, a) = \frac{p(p^{a+1} - 1)}{p^{a+1}(p - 1)}$$

$$\therefore \xi(p, a) = \frac{p}{(p-1)} \times \frac{p^{a+1} - 1}{p^{a+1}}$$

$$\therefore \xi(p, a) = (1 - 1/p^{a+1}) \frac{p}{(p-1)}$$

We have $1 - 1/p^{a+1} < 1$, therefore we observe the following.

$$\xi(p, a) < \frac{p}{(p-1)}$$

This concludes the proof of Lemma 37.

□

5. Induction on prime factorizations

Let m be a positive integer.

Let q be one of the following values.

$q = 2^a$ for $a \geq 5$,

$q = 3^a$ for $a \geq 3$,

$q = 5^a$ for $a \geq 3$,

$q = 7^a$ for $a \geq 2$,

$q = p^a$ for a prime $p \geq 11$ and $a \geq 1$.

We prove in this section that if $n = m \times q$ then $\sigma(n) < \phi(n)$.

Given the prime factorization $m = b_1 \ldots b_k$, we prove by induction on the length k of the factorization that if $\sigma(q) < \phi(q)$ and $n = m \times q$, then $\sigma(n) < \phi(n)$.

The relations $\sigma(q) < \phi(q)$ are proved in Lemmas [38], [39], [40], [41], [42].

The base cases for the induction are proved in Lemmas [43], [44], [45], [46], [48].

The full induction is proved in Lemmas [49], [50], [51], [52], [53].

Lemma 38. Let $q = 2^a$ for $a \geq 5$. Then we observe the following.

$$\sigma(q) < \phi(q)$$

Proof

We need to show $\sigma(q) < \phi(q)$. Using Lemma 36, the following are equivalent.

$$\sigma(q) < \phi(q)$$
$$\sigma(2^a) < \phi(2^a)$$
$$\xi(2^a) < \varphi(2^a)$$

Using Lemma 37, we have the following.

$$\xi(2^a) < 2$$

And for $a > 4$ from Lemma 22, we have the following.

$$2 < \varphi(2^a)$$

Therefore we have the following.

$$\xi(2^a) < 2 < \varphi(2^a)$$

$$\therefore \xi(2^a) < \varphi(2^a)$$

$$\therefore \sigma(2^a) < \phi(2^a)$$

$$\therefore \sigma(q) < \phi(q)$$

This concludes the proof of Lemma 38.

$\square$

Lemma 39. *Let $q = 3^a$ for $a \geq 3$. Then we observe the following.*

$$\sigma(q) < \phi(q)$$

Proof

We need to show $\sigma(q) < \phi(q)$. Using Lemma 36, the following are equivalent.

$$\sigma(q) < \phi(q)$$
$$\sigma(3^a) < \phi(3^a)$$
$$\xi(3^a) < \varphi(3^a)$$

Using Lemma 37, we have the following.

$$\xi(3^a) < 3/2$$

And for $a > 2$ from Lemma 23, we have the following.

$$3/2 < \varphi(3^a)$$

Therefore we have the following.

$$\xi(3^a) < 3/2 < \varphi(3^a)$$

$$\therefore \xi(3^a) < \varphi(3^a)$$

$$\therefore \sigma(3^a) < \phi(3^a)$$

$$\therefore \sigma(q) < \phi(q)$$

This concludes the proof of Lemma 39.

□

Lemma 40. *Let $q = 5^a$ for $a \geq 3$. Then we observe the following.*

$$\sigma(q) < \phi(q)$$

Proof

We need to show $\sigma(q) < \phi(q)$. Using Lemma 36, the following are equivalent.

$$\sigma(q) < \phi(q)$$
$$\sigma(5^a) < \phi(5^a)$$
$$\xi(5^a) < \varphi(5^a)$$

Using Lemma 37, we have the following.

$$\xi(5^a) < 5/4$$

And for $a > 2$ from Lemma 24, we have the following.

$$5/4 < \varphi(5^a)$$

Therefore we have the following.

$$\xi(5^a) < 5/4 < \varphi(5^a)$$

$$\therefore \xi(5^a) < \varphi(5^a)$$

$$\therefore \sigma(5^a) < \phi(5^a)$$

$$\therefore \sigma(q) < \phi(q)$$

This concludes the proof of Lemma 40.

□

Lemma 41. *Let $q = 7^a$ for $a \geq 2$. Then we observe the following.*

$$\sigma(q) < \phi(q)$$

Proof

We need to show $\sigma(q) < \phi(q)$. Using Lemma 36, the following are equivalent.

$$\sigma(q) < \phi(q)$$
$$\sigma(7^a) < \phi(7^a)$$
$$\xi(7^a) < \varphi(7^a)$$

Using Lemma 37, we have the following.

$$\xi(7^a) < 7/6$$

And for $a > 2$ from Lemma 25, we have the following.

$$7/6 < \varphi(7^a)$$

Therefore we have the following.

$$\xi(7^a) < 7/6 < \varphi(7^a)$$

$$\therefore \xi(7^a) < \varphi(7^a)$$

$$\therefore \sigma(7^a) < \phi(7^a)$$

$$\therefore \sigma(q) < \phi(q)$$

This concludes the proof of Lemma 41.

□

Lemma 42. *Let $q = 11^a$ for $a \geq 1$. Then we observe the following.*

$$\sigma(q) < \phi(q)$$

Proof

We need to show $\sigma(q) < \phi(q)$. Using Lemma 36, the following are equivalent.

$$\sigma(q) < \phi(q)$$
$$\sigma(11^a) < \phi(11^a)$$
$$\xi(11^a) < \varphi(11^a)$$

Using Lemma 37, we have the following.

$$\xi(11^a) < 11/10$$

And for $a \geq 1$ from Lemma 26, we have the following.

$$11/10 < \varphi(11^a)$$

Therefore we have the following.

$$\xi(11^a) < 11/10 < \varphi(11^a)$$

$$\therefore \xi(11^a) < \varphi(11^a)$$

$$\therefore \sigma(11^a) < \phi(11^a)$$

$$\therefore \sigma(q) < \phi(q)$$

This concludes the proof of Lemma 42.

□

Lemma 43. *Let $q = 2^a$ for $a \geq 5$. Let b be any prime with $b > 2$. Let $m = q \times b$. We observe the following.*

$$\sigma(m) < \phi(m)$$

Proof

We need to show $\sigma(m) < \phi(m)$, so we need to show the following.

$$\sigma(2^a \times b) < \phi(2^a \times b)$$

Using Lemma 27., Lemma 28. and Lemma 37. the following are equivalent.

$$\sigma(2^a \times b) < \phi(2^a \times b)$$

$$(b+1)\sigma(2^a) < \delta 2^a b \log\log 2^a b$$

$$(b+1)\tau(2,a) < \delta 2^a b \log\log 2^a b$$

$$(b+1) \times \frac{\tau(2,a)}{2^a} < \delta b \log\log 2^a b$$

$$(1 + 1/b) \times \xi(2,a) < \varphi(2^a b)$$

We observe from Lemma 37. the following.

$$(1 + 1/b) \times \xi(2,a) < (1 + 1/b) \times 2$$

Using Lemma 20. the function $f(n) = 1 + 1/n$ is decreasing, hence for $n \geq 3$ we have $f(3) > f(n)$.

Therefore for $a \geq 5$ and $b \geq 3$ the following hold.

$$1 + 1/3 > 1 + 1/b$$

Using Lemma 19, we have that $\varphi(n)$ is strictly increasing. Therefore for $a \geq 5$ and $b \geq 3$ we observe the following numerically.

$$(1 + 1/3) \times 2 < \varphi(2^a b)$$

$$\therefore (1 + 1/b) \times 2 < \varphi(2^a b)$$

We have the following.

$$(1 + 1/b) \times \xi(2, a) < (1 + 1/3) \times 2 < \varphi(2^a b)$$

$$\therefore (1 + 1/b) \times \xi(2, a) < \delta \log \log 2^a b$$

$$\therefore (b + 1)\sigma(2^a) < \delta 2^a b \log \log 2^a b$$

$$\therefore \sigma(2^a b) < \phi(2^a b)$$

$$\therefore \sigma(m) < \phi(m)$$

This concludes the proof of Lemma 43.

□

Lemma 44. *Let $q = 3^a$ for $a \geq 3$. Let b be any prime with $b > 3$. Let $m = q \times b$. We observe the following.*

$$\sigma(m) < \phi(m)$$

Proof

We need to show $\sigma(m) < \phi(m)$, so we need to show the following.

$$\sigma(3^a \times b) < \phi(3^a \times b)$$

Using Lemma 27, and Lemma 28, the following are equivalent.

$$\sigma(3^a \times b) < \phi(3^a \times b)$$

$$(b+1)\sigma(3^a) < \delta 3^a b \log\log 3^a b$$

$$(b+1)\tau(3,a) < \delta 3^a b \log\log 3^a b$$

$$(b+1) \times \frac{\tau(3,a)}{3^a} < \delta b \log\log 3^a b$$

$$(1 + 1/b) \times \xi(3,a) < \delta \log\log 3^a b$$

Using Lemma 20, the function $f(n) = 1 + 1/n$ is decreasing, hence for $n \geq 5$ we have $f(5) > f(n)$.

Therefore for $a \geq 3$ and $b \geq 5$ the following hold.

$$1 + 1/5 > 1 + 1/b$$

Using Lemma 19, we have that $\varphi(n)$ is strictly increasing. Therefore for $a \geq 3$ and $b \geq 5$ we observe the following numerically.

$$(1 + 1/5) \times 3/2 < \varphi(3^a b)$$

$$\therefore (1 + 1/b) \times 3/2 < \varphi(3^a b)$$

We have the following.

$$(1 + 1/b) \times \xi(3, a) < (1 + 1/5) \times 3/2 < \varphi(3^a b)$$

$$\therefore (1 + 1/b) \times \xi(3, a) < \delta \log \log 3^a b$$

$$\therefore (b + 1)\sigma(3^a) < \delta 3^a b \log \log 3^a b$$

$$\therefore \sigma(3^a b) < \phi(3^a b)$$

$$\therefore \sigma(m) < \phi(m)$$

This concludes the proof of Lemma 44.

$\square$

Lemma 45. *Let $q = 5^a$ for $a \geq 3$. Let b be any prime with $b > 5$. Let $m = q \times b$. We observe the following.*

$$\sigma(m) < \phi(m)$$

Proof

We need to show $\sigma(m) < \phi(m)$, so we need to show the following.

$$\sigma(5^a \times b) < \phi(5^a \times b)$$

Using Lemma 27, and Lemma 28, the following are equivalent.

$$\sigma(5^a \times b) < \phi(5^a \times b)$$

$$(b+1)\sigma(5^a) < \delta 5^a b \log \log 5^a b$$

$$(b+1)\tau(5,a) < \delta 5^a b \log \log 5^a b$$

$$(b+1) \times \frac{\tau(5,a)}{5^a} < \delta b \log \log 5^a b$$

$$(1 + 1/b) \times \xi(5,a) < \delta \log \log 5^a b$$

Using Lemma 20, the function $f(n) = 1 + 1/n$ is decreasing, hence for $n \geq 7$ we have $f(7) > f(n)$.

Therefore for $a \geq 3$ and $b \geq 7$ the following hold.

$$1 + 1/7 > 1 + 1/b$$

Using Lemma 19, we have that $\varphi(n)$ is strictly increasing. Therefore for $a \geq 3$ and $b \geq 7$ we observe the following numerically.

$$(1 + 1/7) \times 5/4 < \varphi(5^a \times 3)$$

$$\therefore (1 + 1/b) \times 5/4 < \varphi(5^a b)$$

We have the following.

$$(1 + 1/b) \times \xi(5, a) < (1 + 1/7) \times 5/4 < \varphi(5^a b)$$

$$\therefore (1 + 1/b) \times \xi(5, a) < \delta \log \log 5^a b$$

$$\therefore (b + 1)\sigma(5^a) < \delta 5^a b \log \log 5^a b$$

$$\therefore \sigma(5^a b) < \phi(5^a b)$$

$$\therefore \sigma(m) < \phi(m)$$

This concludes the proof of Lemma 45.

□

Lemma 46. Let $q = 7^a$ for $a \geq 2$. Let b be any prime with $b > 7$. Let $m = q \times b$. We observe the following.

$$\sigma(m) < \phi(m)$$

Proof

We need to show $\sigma(m) < \phi(m)$, so we need to show the following.

$$\sigma(7^a \times b) < \phi(7^a \times b)$$

Using Lemma 27, and Lemma 28, the following are equivalent.

$$\sigma(7^a \times b) < \phi(7^a \times b)$$

$$(b+1)\sigma(7^a) < \delta 7^a b \log \log 7^a b$$

$$(b+1)\tau(7, a) < \delta 7^a b \log \log 7^a b$$

$$(b+1) \times \frac{\tau(7, a)}{7^a} < \delta b \log \log 7^a b$$

$$(1 + 1/b) \times \xi(7, a) < \delta \log \log 7^a b$$

Using Lemma 20, the function $f(n) = 1 + 1/n$ is decreasing, hence for $n \geq 11$ we have $f(11) > f(n)$.

Therefore for $a \geq 2$ and $b \geq 11$ the following hold.

$$1 + 1/11 > 1 + 1/b$$

Using Lemma 19, we have that $\varphi(n)$ is strictly increasing. Therefore for $a \geq 3$ and $b \geq 11$ we observe the following numerically.

$$(1 + 1/11) \times 7/6 < \varphi(11^a b)$$

$$\therefore (1 + 1/b) \times 7/6 < \varphi(11^a b)$$

We have the following.

$$(1 + 1/b) \times \xi(7, a) < (1 + 1/11) \times 3/2 < \varphi(7^a b)$$

$$\therefore (1 + 1/b) \times \xi(7, a) < \delta \log \log 7^a b$$

$$\therefore (b+1)\sigma(7^a) < \delta 7^a b \log \log 7^a b$$

$$\therefore \sigma(7^a b) < \phi(7^a b)$$

$$\therefore \sigma(m) < \phi(m)$$

This concludes the proof of Lemma 46.

□

Lemma 47. *Let b be a prime with $b \geq 11$. Let $m = b^a$ for $a \geq 1$.*

We observe the following.

$$\sigma(m) < \phi(m)$$

Proof

We need to show $\sigma(m) < \phi(m)$, so we need to show the following.

$$\sigma(b^a) < \phi(b^a)$$

Proceed by induction on a.

If $a = 1$ we have from Lemma 21, that $\sigma(b) < \phi(b)$.

Suppose $\sigma(b^a) < \phi(b^a)$ for some a. Consider the case $a+1$.

Using Lemma 34, we have the following.

$$\sigma(b^{a+1}) \leq (b+1)\sigma(b^a)$$

From the inductive hypothesis we have the following.

$$(b+1)\sigma(b^a) < (b+1)\phi(b^a)$$

The following are equivalent.

$$(b+1)\phi(b^a) < \phi(b^{a+1})$$

$$(b+1)\delta b^a \log \log b^a < \delta b^{a+1} \log \log b^{a+1}$$

$$(b+1) \log \log b^a < b \log \log b^{a+1}$$

$$(1 + 1/b) \log \log b^a < \log \log b^{a+1}$$

Using Lemma 20, the function $f(n) = 1 + 1/n$ is decreasing, hence for $n \geq 11$ we have $f(11) \geq f(n)$.

Therefore for $b \geq 11$ the following holds.

$$1 + 1/11 \geq 1 + 1/b$$

We have the following.

$$(1 + 1/b) \log \log b^a \leq (1 + 1/11) \log \log b^a$$

The function $f(n) = \log \log n$ is increasing. For $b = 11$ and $a = 1$ we observe the following.

$$(1 + 1/11) \log \log 11 < \log \log 11^2$$

$$\therefore (1 + 1/11) \log \log 11^a < \log \log 11^{a+1}$$

$$\therefore (1 + 1/11) \log \log b^a < \log \log b^{a+1}$$

$$\therefore (1 + 1/b) \log \log b^a < \log \log b^{a+1}$$

$$\therefore (b+1)\phi(b^a) < \phi(b^{a+1})$$

$$\therefore \sigma(b^{a+1}) < \phi(b^{a+1})$$

By induction, $\sigma(b^a) < \phi(b^a)$. Therefore $\sigma(m) < \phi(m)$.

This concludes the proof of Lemma 47.

□

Lemma 48. *Let p be a prime with $p \geq 11$. Let $q = p^a$ for $a \geq 1$. Let b be a prime with $b \geq 11$. Let $m = q \times b$. We observe the following.*

$$\sigma(m) < \phi(m)$$

Proof

We need to show $\sigma(m) < \phi(m)$, so we need to show $\sigma(qb) < \phi(qb)$.

Using Lemma 47, we have $\sigma(q) < \phi(q)$.

Suppose $b \nmid q$. Using Lemma 30, and Lemma 31, the following are equivalent.

$$\sigma(q) < \phi(q)$$

$$\Lambda(q) < \phi(q)$$

$$(b+1)\Lambda(q) < \phi(qb)$$

$$\sigma(qb) < \phi(qb)$$

Suppose $b \mid q$. Using Lemma 30, Lemma 29, and Lemma 33, we have the following.

$$\sigma(qb) = b\Lambda(q) + \Lambda(\lambda(q \setminus b))$$

$$\sigma(qb) \leq b\sigma(q) + \Lambda(q)$$

$$\sigma(qb) \leq (b+1)\sigma(q)$$

$$\sigma(qb) < (b+1)\phi(q) < \phi(qb)$$

$$\therefore \sigma(qb) < \phi(qb)$$

$$\therefore \sigma(m) < \phi(m)$$

This concludes the proof of Lemma 48.

$\square$

Lemma 49. *Let m be a positive integer. Let $q = 2^a$ with $a \geq 5$. Let $n = m \times q$. We observe the following.*

$$\sigma(n) < \phi(n)$$

Proof

We need to show $\sigma(n) < \phi(n)$ so we need to show $\sigma(mq) < \phi(mq)$.

Write $m = b_1 \ldots b_k$. Proceed by induction on k.

If $k = 1$ then $m = b_1$. Using Lemma 43, we have the following.

$$\sigma(b_1 q) < \phi(b_1 q)$$

Suppose for some k we have the following.

$$\sigma(b_1 \ldots b_k q) < \phi(b_1 \ldots b_k q)$$

Assume without loss of generality that $b_i \neq 2$ for all i with $1 \leq i \leq k$.

$$\therefore \sigma(2^a b_1 \ldots b_k) < \phi(2^a b_1 \ldots b_k)$$

$$\therefore \sigma(2^a b_1 \ldots b_k) < \delta \times 2^a b_1 \ldots b_k \log \log(2^a b_1 \ldots b_k)$$

Consider the case of $k+1$. We need to show the following.

$$\sigma(2^a b_1 \ldots b_k b_{k+1}) < \phi(2^a b_1 \ldots b_k b_{k+1})$$

Using Lemma 30, we have

$$\sigma(2^a b_1 \ldots b_k b_{k+1}) = \Lambda(2^a b_1 \ldots b_k b_{k+1})$$

Using Lemma 34, we have $\sigma(2^a b_1 \ldots b_k b_{k+1}) \leq (b_{k+1}+1)\sigma(m)$.

Using the inductive hypothesis we have the following.

$$(b_{k+1}+1)\sigma(2^a b_1 \ldots b_k) < (b_{k+1}+1)\delta \times 2^a b_1 \ldots b_k \log\log(2^a b_1 \ldots b_k)$$

$$\therefore \sigma(2^a b_1 \ldots b_k b_{k+1}) < (b_{k+1}+1)\delta \times 2^a b_1 \ldots b_k \log\log(2^a b_1 \ldots b_k)$$

Using Lemma 29, the following holds.

$$(b_{k+1}+1)\phi(2^a b_1 \ldots b_k) < \phi(2^a b_1 \ldots b_k b_{k+1})$$

Therefore we have the following.

$$\sigma(2^a b_1 \ldots b_k b_{k+1}) < \phi(2^a b_1 \ldots b_k b_{k+1})$$

Therefore by induction $\sigma(mq) < \phi(mq)$. Therefore $\sigma(n) < \phi(n)$.

This concludes the proof of Lemma 49.

□

Lemma 50. *Let m be a positive integer. Let $q = 3^a$ with $a \geq 3$. Let $n = m \times q$. We observe the following.*

$$\sigma(n) < \phi(n)$$

Proof

We need to show $\sigma(n) < \phi(n)$ so we need to show $\sigma(mq) < \phi(mq)$.

Write $m = b_1 \ldots b_k$. Proceed by induction on k.

If $k = 1$ then $m = b_1$. Using Lemma 44, we have the following.

$$\sigma(b_1 q) < \phi(b_1 q)$$

Suppose for some k we have the following.

$$\sigma(b_1 \ldots b_k q) < \phi(b_1 \ldots b_k q)$$

Due to Lemma 49, we may assume without loss of generality that $b_i \neq 2$ for all i with $1 \leq i \leq k$.

Since $3^a \times 3 = 3^{a+1}$ we may assume without loss of generality that $b_i \neq 3$ for all i with $1 \leq i \leq k$.

$$\therefore \sigma(3^a b_1 \ldots b_k) < \phi(3^a b_1 \ldots b_k)$$

$$\therefore \sigma(3^a b_1 \ldots b_k) < \delta \times 3^a b_1 \ldots b_k \log\log(3^a b_1 \ldots b_k)$$

Consider the case of $k + 1$. We need to show the following.

$$\sigma(3^a b_1 \ldots b_k b_{k+1}) < \phi(3^a b_1 \ldots b_k b_{k+1} q)$$

Using Lemma 30, we have the following.

$$\sigma(3^a b_1 \ldots b_k b_{k+1}) = \Lambda(3^a b_1 \ldots b_k b_{k+1})$$

Using Lemma 34, we have $\sigma(3^a b_1 \ldots b_k b_{k+1}) \leq (b_{k+1} + 1)\sigma(m)$.

Using the inductive hypothesis we have the following.

$$(b_{k+1} + 1)\sigma(3^a b_1 \ldots b_k) < (b_{k+1} + 1)\delta \times 3^a b_1 \ldots b_k \log\log(3^a b_1 \ldots b_k)$$

$$\therefore \sigma(3^a b_1 \ldots b_k b_{k+1}) < (b_{k+1} + 1)\delta \times 3^a b_1 \ldots b_k \log\log(3^a b_1 \ldots b_k)$$

Using Lemma 29, the following holds.

$$(b_{k+1} + 1)\phi(3^a b_1 \ldots b_k) < \phi(3^a b_1 \ldots b_k b_{k+1})$$

Therefore we have the following.

$$\sigma(3^a b_1 \ldots b_k b_{k+1}) < \phi(3^a b_1 \ldots b_k b_{k+1})$$

Therefore by induction $\sigma(mq) < \phi(mq)$. Therefore $\sigma(n) < \phi(n)$.

This concludes the proof of Lemma 50.

□

Lemma 51. *Let m be a positive integer. Let $q = 5^a$ with $a \geq 3$. Let $n = m \times q$. We observe the following.*

$$\sigma(n) < \phi(n)$$

Proof

We need to show $\sigma(n) < \phi(n)$ so we need to show $\sigma(mq) < \phi(mq)$.

Write $m = b_1 \ldots b_k$. Proceed by induction on k.

If $k = 1$ then $m = b_1$. Using Lemma 45, we have the following.

$$\sigma(b_1 q) < \phi(b_1 q)$$

Suppose for some k we have the following.

$$\sigma(b_1 \ldots b_k q) < \phi(b_1 \ldots b_k q)$$

Due to Lemma 49, we may assume without loss of generality that $b_i \neq 2$ for all i with $1 \leq i \leq k$.

Due to Lemma 50, we may assume without loss of generality that $b_i \neq 2$ for all i with $1 \leq i \leq k$.

Since $5^a \times 5 = 5^{a+1}$ we may assume without loss of generality that $b_i \neq 5$ for all i with $1 \leq i \leq k$.

$$\therefore \sigma(5^a b_1 \ldots b_k) < \phi(5^a b_1 \ldots b_k)$$

$$\therefore \sigma(5^a b_1 \ldots b_k) < \delta \times 5^a b_1 \ldots b_k \log\log(5^a b_1 \ldots b_k)$$

Consider the case of $k + 1$. We need to show the following.

$$\sigma(5^a b_1 \ldots b_k b_{k+1}) < \phi(5^a b_1 \ldots b_k b_{k+1})$$

Using Lemma 30, we have the following.

$$\sigma(5^a b_1 \ldots b_k b_{k+1}) = \Lambda(5^a b_1 \ldots b_k b_{k+1})$$

Using Lemma 34, we have $\sigma(5^a b_1 \ldots b_k b_{k+1}) \leq (b_{k+1} + 1)\sigma(m)$.

Using the inductive hypothesis we have the following.

$$(b_{k+1} + 1)\sigma(5^a b_1 \ldots b_k) < (b_{k+1} + 1)\delta \times 5^a b_1 \ldots b_k \log\log(5^a b_1 \ldots b_k)$$

$$\therefore \sigma(5^a b_1 \ldots b_k b_{k+1}) < (b_{k+1} + 1)\delta \times 5^a b_1 \ldots b_k \log\log(5^a b_1 \ldots b_k)$$

Using Lemma 29, the following holds.

$$(b_{k+1} + 1)\phi(5^a b_1 \ldots b_k) < \phi(5^a b_1 \ldots b_k b_{k+1})$$

Therefore we have the following.

$$\sigma(5^a b_1 \ldots b_k b_{k+1}) < \phi(5^a b_1 \ldots b_k b_{k+1})$$

Therefore by induction $\sigma(mq) < \phi(mq)$. Therefore $\sigma(n) < \phi(n)$.

This concludes the proof of Lemma 51.

Lemma 52. *Let m be a positive integer. Let $q = 7^a$ with $a \geq 2$. Let $n = m \times q$. We observe the following.*

$$\sigma(n) < \phi(n)$$

Proof

We need to show $\sigma(n) < \phi(n)$ so we need to show $\sigma(mq) < \phi(mq)$.

Write $m = b_1 \ldots b_k$. Proceed by induction on k.

If $k = 1$ then $m = b_1$. Using Lemma 46, we have the following.

$$\sigma(b_1 q) < \phi(b_1 q)$$

Suppose for some k we have the following.

$$\sigma(b_1 \ldots b_k q) < \phi(b_1 \ldots b_k q)$$

Due to Lemma 49, we may assume without loss of generality that $b_i \neq 2$ for all i with $1 \leq i \leq k$.

Due to Lemma 50, we may assume without loss of generality that $b_i \neq 3$ for all i with $1 \leq i \leq k$.

Due to Lemma 51, we may assume without loss of generality that $b_i \neq 5$ for all i with $1 \leq i \leq k$.

Since $7^a \times 7 = 7^{a+1}$ we may assume without loss of generality that $b_i \neq 7$ for all i with $1 \leq i \leq k$.

$$\therefore \sigma(7^a b_1 \ldots b_k) < \phi(7^a b_1 \ldots b_k)$$

$$\therefore \sigma(7^a b_1 \ldots b_k) < \delta \times 7^a b_1 \ldots b_k \log\log(7^a b_1 \ldots b_k)$$

Consider the case of $k + 1$. We need to show the following.

$$\sigma(7^a b_1 \ldots b_k b_{k+1}) < \phi(7^a b_1 \ldots b_k b_{k+1})$$

Using Lemma 30, we have the following.

$$\sigma(7^a b_1 \ldots b_k b_{k+1}) = \Lambda(7^a b_1 \ldots b_k b_{k+1})$$

Using Lemma 34, we have $\sigma(7^a b_1 \ldots b_k b_{k+1}) \leq (b_{k+1} + 1)\sigma(m)$.

Using the inductive hypothesis we have the following.

$$(b_{k+1} + 1)\sigma(7^a b_1 \ldots b_k) < (b_{k+1} + 1)\delta \times 7^a b_1 \ldots b_k \log\log(7^a b_1 \ldots b_k)$$

$$\therefore \sigma(7^a b_1 \ldots b_k b_{k+1}) < (b_{k+1} + 1)\delta \times 7^a b_1 \ldots b_k \log\log(7^a b_1 \ldots b_k)$$

Using Lemma 29, the following holds.

$$(b_{k+1} + 1)\phi(7^a b_1 \ldots b_k) < \phi(7^a b_1 \ldots b_k b_{k+1})$$

Therefore we have the following.

$$\sigma(7^a b_1 \ldots b_k b_{k+1}) < \phi(7^a b_1 \ldots b_k b_{k+1})$$

Therefore by induction $\sigma(mq) < \phi(mq)$. Therefore $\sigma(n) < \phi(n)$.

This concludes the proof of Lemma 52.

□

Lemma 53. *Let m be a positive integer. Let p be a prime with $p \geq 11$. Let $q = p^a$ with $a \geq 1$. Let $n = m \times p$. We observe the following.*

$$\sigma(n) < \phi(n)$$

Proof

We need to show $\sigma(n) < \phi(n)$ so we need to show $\sigma(mq) < \phi(mq)$.

Write $m = b_1 \ldots b_k$. Proceed by induction on k.

If $k = 1$ then $m = b_1$. Using Lemma 48, we have the following.

$$\sigma(b_1 q) < \phi(b_1 q)$$

Suppose for some k we have the following.

$$\sigma(b_1 \ldots b_k q) < \phi(b_1 \ldots b_k q)$$

Due to Lemma 49, Lemma 50, Lemma 51 and Lemma 52, we may assume without loss of generality that $b_i \neq 2, 3, 5, 7$.

Since $p^a \times p = p^{a+1}$ we may assume without loss of generality that $b_i \neq p$ for all i with $1 \leq i \leq k$.

$$\therefore \sigma(p^a b_1 \ldots b_k) < \phi(p^a b_1 \ldots b_k)$$

$$\therefore \sigma(p^a b_1 \ldots b_k) < \delta \times p^a b_1 \ldots b_k \log\log(p^a b_1 \ldots b_k)$$

Consider the case of $k+1$. We need to show the following.

$$\sigma(p^a b_1 \ldots b_k b_{k+1}) < \phi(p^a b_1 \ldots b_k b_{k+1} n)$$

Using Lemma 30, we have the following.

$$\sigma(p^a b_1 \ldots b_k b_{k+1}) = \Lambda(p^a b_1 \ldots b_k b_{k+1})$$

Using Lemma 34, we have $\sigma(p^a b_1 \ldots b_k b_{k+1}) \leq (b_{k+1} + 1)\sigma(m)$.

Using the inductive hypothesis we have the following.

$$(b_{k+1} + 1)\sigma(p^a b_1 \ldots b_k) < (b_{k+1} + 1)\delta \times p^a b_1 \ldots b_k \log\log(p^a b_1 \ldots b_k)$$

$$\therefore \sigma(p^a b_1 \ldots b_k b_{k+1}) < (b_{k+1} + 1)\delta \times p^a b_1 \ldots b_k \log\log(p^a b_1 \ldots b_k)$$

Since $p \geq 11$ and $b_{k+1} \geq 11$ using Lemma 29, the following holds.

$$(b_{k+1} + 1)\phi(p^a b_1 \ldots b_k) < \phi(p^a b_1 \ldots b_k b_{k+1})$$

Therefore we have the following.

$$\sigma(p^a b_1 \ldots b_k b_{k+1}) < \phi(p^a b_1 \ldots b_k b_{k+1})$$

Therefore by induction $\sigma(mq) < \phi(mq)$. Therefore $\sigma(n) < \phi(n)$.

This concludes the proof of Lemma 53.

□

6. Robin's condition holds

From Remark 13. Robin's sufficient condition for the Riemann Hypothesis may be described as follows.

If the value of the sum of divisors function $\sigma(m)$ exceeds $e^\gamma m \log\log m$ finitely many times only, then the Riemann Hypothesis holds.

We prove in this section that the inequality $\sigma(n) < e^\gamma n \log\log n$ holds for all $n \geq \Omega$. Hence $\sigma(n) \geq e^\gamma n \log\log n$ only for a finite number of values for n.

Lemma 54. states that if $n \geq \Omega$ then n can be written as a product of the form $n = q \times m$ for some m and q a power of a prime.

Theorem 55. asserts that if $n \geq \Omega$ then $\sigma(n) < \phi(n)$.

Corollary 56. is the final result and states that $\sigma(n) \geq e^\gamma n \log\log n$ for only a finite number of values for n.

Lemma 54. *Let $n \geq \Omega$. Let $q = \alpha(n)$. Then $n = q \times m$ for some m.*

Proof

Let $q = \alpha(n)$ and let $m = \beta(n)$.

From Lemma 18, we have that q is a factor of n.

The values of $\alpha(n)$ are as follows.

$$\alpha(n) = 2^5 \text{ for } n = 2^a \text{ with } a \geq 5,$$

$$\alpha(n) = 3^3 \text{ for } n = 3^a \text{ with } a \geq 3,$$

$$\alpha(n) = 5^3 \text{ for } n = 5^a \text{ with } a \geq 3,$$

$$\alpha(n) = 7^2 \text{ for } n = 7^a \text{ with } a \geq 2,$$

$$\alpha(n) = n \text{ for } n = b^a \text{ with } b \geq 11 \text{ and } a \geq 1.$$

The value of $\beta(n)$ is $\beta(n) = n/\alpha(n)$.

Note both $\alpha(n)$ and $\beta(n)$ are of integral value.

We have the following.

$$n = \alpha(n) \times n/\alpha(n)$$

$$\therefore n = \alpha(n) \times \beta(n) = \alpha(n) \times m$$

$$\therefore n = \alpha(n) \times m$$

This concludes the proof of Lemma 54.

□

Theorem 55. *For $n \geq \Omega$ we have $\sigma(n) < \phi(n)$.*

Proof

Let $n \geq \Omega$. Using Lemma 18, we have that either one of the following holds.

$2^a \mid n$ for $a \geq 5$,

$3^a \mid n$ for $a \geq 3$,

$5^a \mid n$ for $a \geq 3$,

$7^a \mid n$ for $a \geq 2$,

$b^a \mid n$ for prime $b \geq 11$ and $a \geq 1$.

Using Lemma 54, we have $n = m \times q$ for some m and $q = \alpha(n)$.

Note $\alpha(n) \geq 11$, so we may apply the results of Lemmas 49, 50, 51, 52, 53.

If $2^a \mid n$ for $a \geq 5$ then $q = 2^5$ and result follows from Lemma 49.

If $3^a \mid n$ for $a \geq 3$ then $q = 3^3$ and result follows from Lemma 50.

If $5^a \mid n$ for $a \geq 3$ then $q = 5^3$ and result follows from Lemma 51.

If $7^a \mid n$ for $a \geq 2$ then $q = 7^2$ and result follows from Lemma 52.

If $b^a \mid n$ for $a \geq 1$ and $b \geq 11$ then $q = b$ and result follows from Lemma 53.

This concludes the proof of Theorem 55.

□

Corollary 56. follows from Definition 6., Remark 5. and Theorem 55.

Corollary 56. $\sigma(m) \geq e^\gamma m \log \log m$ *for finitely many m.*

7. Conclusion

For a positive integer n with $n \geq \Omega$, we have proved that $\sigma(n) < \phi(n)$.

It is easy to see that $n = \alpha(n) \times \beta(n)$.

$$\alpha(n) \times \beta(n) = \alpha(n) \times n/\alpha(n) = n$$

Both $\alpha(n)$ and $\beta(n)$ are of integral value. Furthermore, $\alpha(n) \geq 11$. This is the content of Lemma 54.

We let $m = \alpha(n)$ and $q = \beta(n)$.

Thus for $n = m \times q$ we obtain in Theorem 55, that $\sigma(n) < \phi(n)$

Thus we have shown that if $n \geq 2^5 \times 3^3 \times 5^3 \times 7^2$ the following holds.

$$\sigma(n) < \phi(n) < \delta n \log \log n < e^\gamma n \log \log n$$

$$\therefore \sigma(n) < e^\gamma n \log \log n$$

Hence $\sigma(n) < e^\gamma n \log \log n$ for $n \geq 2^5 \times 3^3 \times 5^3 \times 7^2$.

This implies that $\sigma(n) \geq e^\gamma n \log \log n$ for only a finite number of values for n.

Acknowledgements

Professor Salzer pointed out one of the formulas I was conjecturing was in fact a well-known result. Some of the notation used above is inspired from our studies on Universal Algebra, focusing on work carried out in the eighties in Chisinau, the capital city of Moldavia.

Acknowledgements are due to the team at HAL Archives-ouvertes for publishing many of my scientific manuscripts, some of them scientifically relevant to the issues our Society is facing today. I have been developing my skills in speaking and writing in French for most of my life now. We are fortunate in this day and age to live in a world in which ideas flow freely. I therefore thank HAL for making the dissemination of my ideas possible.

I thank my mother Raluca Bura for her support along the years, and the people at her place of work Saint John of God Villa, in Subiaco.

I am highly indebted to Noam Greenberg for a research-intensive year of 2012, at Victoria University of Wellington in New Zealand.

Between the years of 2010 and 2013 Victoria University of Wellington provided a good environment for research, and I experienced excitement and intellectual stimulation studying in their Departments of Mathematics and Philosophy.

I further thank all of those from which I have learned.

I thank the following people for their encouragement, academic friendship or involvement in my research. Noam Greenberg, Jeremy Seligman, Cristian Calude, Joshua Arulanandham, David MacIntyre, Max Cresswell, Ed Mares, BD Kim, Rodney Downey, Rob Goldblatt, Peter Donelan, Colin Bailey, Carolyn Chun, David Diamondstone, Asher Kach, Dan Turetzky, Chris Atkin, Dillon Mayhew, Gernot Salzer, Miki Hermann, Mark Reynolds, Tim French, Igor Potapov.

Perth, Western Australia
November 2020

References

[1] G. Robin, *Grandes valeurs de la fonction somme des diviseurs et hypothese de Riemann.* J. Math. pures appl. vol. 63 no. 2 1984.

[2] B. Riemann, *Ueber die Anzahl der Primzahlen unter einer gegebenen Grosse.* Ges. Math. Werke und Wissenschaftlicher Nachlaß vol 2. 1859.

[3] L. Alaoglu and P. Erdős, *On highly composite and similar numbers.* Transactions of the American Mathematical Society. vol. 56 no. 3 1944.

[4] L. Mirsky and P. Erdős, *The distribution of values of the divisor function $d(n)$.* Proceedings of the London Mathematical Society. vol. 2 no. 3 1952.

[5] G. H. Hardy and E. M. Wright, *An introduction to the theory of numbers.* Oxford university press. 1979.

[6] S. Ramanujan, *Highly composite numbers.* Proceedings of the London Mathematical Society. vol. 2 no. 1 1915.

[7] U. Balakrishnan, *On the sum of divisors function.* Journal of Number Theory. vol. 51 no. 2 1995.

[8] D.R. Heath-Brown, *The divisor function at consecutive integers.* Mathematika. vol.31 no. 1 1984.

[9] G. Coppola and S. Salerno, *On the symmetry of the divisor function in almost all short intervals.* Acta Arithmetica. vol. 113 2004.

[10] F. Luca and I. E. Shparlinski, *On the values of the divisor function.* Monatshefte für Mathematik. vol. 154 no. 1 Springer. 2008.

[11] L. Matthiesen, *Correlations of the divisor function.* Proceedings of the London Mathematical Society. vol. 104 no. 4 2012.

[12] P. Pollack and C. Pomerance, *Some problems of Erdős on the sum-of-divisors function.* Transactions of the American Mathematical Society, vol.3 no. 1 2016.

www.ingramcontent.com/pod-product-compliance
Lightning Source LLC
Chambersburg PA
CBHW080001230526
45470CB00008B/2822